Design Thinking:
Practicing Design Competence

Denny C. Davis, PhD
Emeritus Professor, Engineering Education
Washington State University

©2018 Verity Design Learning

Mascoutah, Illinois

http://VerityDesignLearning.com

Shady_oaks@frontier.com

All rights reserved. Contents of this book are intended for use by their owner, not to be copied and distributed to multiple users.

DESIGN THINKING

A Toolbox for Attaining Authentic Design Competence

If you are part of a design project team, this book is for you. Your project faces delays, cost overruns, and damaged reputations if you make mistakes or miss the obvious. As you progress through your design effort, this book helps your team think critically and creatively to avoid or detect problems and to find paths forward that add real value. Reflective design thinking builds authentic design competence.

This book is crafted to prompt team discussions around important steps in the design process. It will also punctuate your efforts with design reviews that help you examine both design process and design products at various stages of development. Design reviews help you judge whether you are ready to advance to the next stage of design, while they also improve your team's ability to explain and critique your work.

Activities in this book are based on industry practices, design research, and decades of guiding students in widely varied team projects. Notes coming from your design discussions and design reviews will be assets for defending your work to clients, filing for patents, and claiming design competence.

Direct your questions or comments to:
Dr. Denny Davis, Verity Design Learning,
Shady_Oaks@frontier.com.

TOOLBOX FOR DESIGN THINKING

Introduction and Purpose

This book is a toolbox for project teams to grow design competence while achieving excellence in design solutions. Pages of this book guide team discussions that challenge assumptions, ideas, and decisions that otherwise may produce an inferior solution. Pages may be used on a pre-planned schedule or just-in-time to fit your team's stage in your project.

Goals

Through these design thinking exercises, you will:

1. Learn to think critically and creatively in design projects
2. Develop and document processes that bring efficiency and credibility to your design efforts
3. Produce solutions that are responsible, offer value, and meet expectations of diverse stakeholders

DESIGN PROJECT STAGES

Stages of Design
Projects must progress from start to completion in an allotted time period. Design thinking will shape the processes and products along the way, advancing through each of three major stages of design shown below. Critical review of <u>processes</u> might best occur near the midpoint of each stage, and reviews of design <u>products</u> at the end of each stage.

1. **Problem Scoping** stage: Gather information to understand the problem; craft a clear definition of problem scope and requirements for the solution.
2. **Concept Generation** stage: Generate and filter ideas for solution parts; select the best, synthesize and refine concepts to pose a plausible solution.
3. **Solution Completion (or Solution Realization)** stage: Analyze, detail, assess risks, test, refine, and validate a final solution with users. Prepare for project hand-off and use.

Project Timeline

Problem Scoping — Concept Generation — Solution Completion

Start → Problem Scoping Review → Concept Generation Review → Solution Completion Review → End

ACTIVITY STRUCTURE

Page Content

Most pages contain these elements:

- ❖ Timing: Symbol marking when to hold the discussion
- ❖ Title: Topic to be addressed in discussion
- ❖ Context: How the topic is relevant
- ❖ Question: Questions to discuss among teammates
- ❖ Principle: A key principle or fact related to the topic
- ❖ Follow-up: Suggestions for acting on what is learned

Facilitation Suggestions

Each page contains a topic for discussion related to your team's current status in your project. The discussion will often focus design activities for that session – potentially surfacing issues that need to be addressed as part of the day's work.

To maximize value from discussions, your team should:

1. Read and think about the activity in advance
2. Assign a facilitator (e.g., team leader) to lead the discussion
3. Draw every member into the discussion to uncover different perspectives
4. Work toward decisions or consensus understandings and record them
5. Document steps identified for follow-up, along with responsible persons and due dates
6. Document learning/discoveries of the team and individuals

Table of Contents

Design Process — 8
    What happens in design? — 9
    What is design iteration? — 10
    Who is needed in design? — 11
    How is time allocated? — 12
    What motivates us in this project? — 13

Problem Definition — 14
    Stakeholders of design — 15
    Understanding project scope — 16
    Defining user needs — 17
    Identifying solution constraints — 18
    Reflection on Process: Information gathering — 19
    Value proposition — 20
    Stakeholder needs — 21
    Project constraints — 22
    Problem definition — 23
    Design Review: Defined problem — 24

Concept Generation — 25
    Generating ideas — 26
    Screening ideas — 27
    Synthesizing concepts — 28
    Selecting best concepts — 29
    Reflection on Process: Concept Selection — 30
    Improving the concept — 31
    Specifying component interfaces — 32
    Specifying functionality — 33
    Specifying costs — 34
    Specifying technical requirements — 35
    Specifying social requirements — 36
    Design Review: Selected Concept — 37

Table of Contents (continued)

Solution Completion/Realization	**38**
Design for prevention	39
Design for efficiency	40
Design for producability	41
Design for sustainability	42
Prototyping for component testing	43
Testing and evaluation	44
Iterating to revisit earlier work	45
Reflection on Process: Solution completion	46
Verifying solution	47
Refining solution	48
Validating solution	49
Design Review: Completed solution	50
Assessment: Design thinking achieved	51
Assessment: Design learning achieved	52

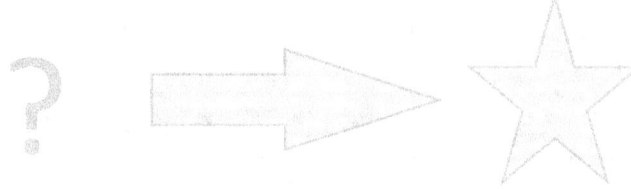

Design Process

Design activities that engage a design team to develop a tangible product, process, or system that meets needs of people or organizations

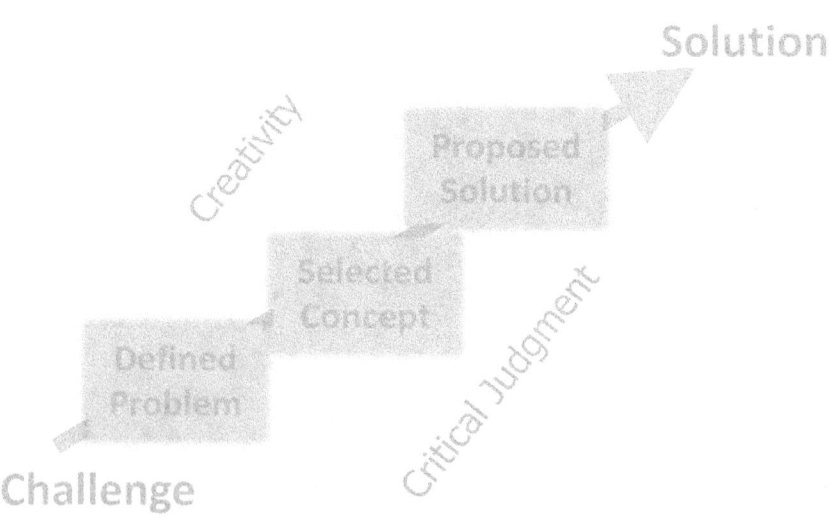

DESIGN PROCESS

| Definition | Conceptualization | Realization |

What Happens in Design?

Overview of Design Process
Design addresses a need or an opportunity for a better solution. As shown below, designers repeatedly apply creativity and analysis to partial solutions before converging on the solution they propose.

Creativity

NEED → Defined Problem → Selected Concept → Preliminary Design → Proposed Solution → SOLUTION

Analysis/Judgment

Design Process
As a team, discuss:
1. How might creativity be applied in design?
2. How might analysis be applied in design?
3. How is the solution affected if creativity or analysis is neglected?

Principle
It is better to give feedback, even when it points out needs for improvement, than to withhold feedback altogether.

Learning and Follow-up
Identify ways to facilitate creative thinking in a team.

Identify tools your team might need to use to analyze the merits of design ideas.

Rate your grasp of this topic. (1=poor, 10=superb)   1 2 3 4 5 6 7 8 9 10

Date completed: Reference: Proverb 27:5

DESIGN PROCESS

Definition	Conceptualization	Realization

What is Design Iteration?

Introduction to Design
Design activity creates technological products which people need or want. Designers work to make their solutions satisfy potential users, investors, industry standards, and the public. Although the design process is depicted below as a linear set of steps (from left to right), designers often repeat earlier steps to correct errors and respond to what is learned along the way. Therefore, design is iterative.

Design Process
As a team, discuss:
1. Why are the three stages of design in the order shown?
2. Under what conditions might a team decide to revisit "problem scoping"?
3. Under what conditions might "concept generation" need to be redone?

Principle
When confronted on a mistake, take heed and you will learn much more in the end.

Learning and Follow-up
List types of problems that might require you to return to an earlier stage of design.

What barriers might make you resist returning to repeat something?

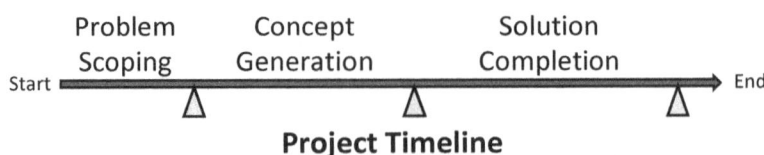

Project Timeline

| Rate your grasp of this topic. (1=poor, 10=superb) | 1 | 2 | 3 | 4 | 5 | 6 | 7 | 8 | 9 | 10 |

Date completed:　　　　　Reference: Proverb 1:23　　　　　10

DESIGN PROCESS

| Definition | Conceptualization | Realization |

Who is Needed in Design?

Demands of Design

Design is much more than tinkering to find a solution to a problem. Design teams need to stay on schedule, find resources, generate ideas, make decisions, solve problems, create parts, integrate parts, document work, and communicate with others. Before starting your project, inventory the abilities and interests of team members so you can capitalize on strengths and find help in areas of need.

Design Personnel

As a team, discuss:
1. Who has what technical skills? Who has strong people skills?
2. Who likes to be creative? Who wants well-defined work?
3. Who has what communication skills? What technologies are available to support your work sharing?
4. What skills or knowledge are lacking? Where is help to be found?

Principle
Be sure that you know the condition of your resources and take care of them.

Learning and Follow-up

Identify for each member the types of project assignments that fit their interests, enable them to learn, and achieve project goals.

What training, self-study, or other help is needed by each member?

Rate your grasp of this topic.
(1=poor, 10=superb)   1 2 3 4 5 6 7 8 9 10

Date completed: Reference: Proverb 27:23

DESIGN PROCESS

| Definition | Conceptualization | Realization |

How is Time Allocated?

Time Demands
Design projects usually have specified end dates, as well as deadlines for intermediate reports and deliverables. Teams must set their own schedules to get their work done on time. In setting schedules, consider the following: time needed for each task, allowable overlap in tasks, and where unexpected delays may occur.

Setting Milestones
As a team, discuss:
1. What fraction of your time can be used for problem scoping? For concept generation? For solution completion?
2. What are major tasks and time demands in problem scoping?
3. What are major tasks and time demands in concept generation?
4. What are major tasks and time demands in solution completion?
5. Where is overlap allowed? Where are potential delays?

Principle
If you don't do what is needed at the proper time, in the end you will have nothing.

Learning and Follow-up
Identify suitable deadlines for each of the following:
- Information gathering
- Defining user needs
- Generating ideas
- Selecting best concept
- Establishing solution requirements
- Prototyping/testing components
- Integrating solution components
- Evaluating solution against requirements
- Refining/testing to improve design
- Documenting final solution

Rate your grasp of this topic.
(1=poor, 10=superb)   1   2   3   4   5   6   7   8   9   10

Date completed: Reference: Proverb 20:4

DESIGN PROCESS

| Definition | Conceptualization | Realization |

What Motivates Us in this Project?

Motivation
People engage in a design project with different motivations. Both intrinsic (internal value-driven) and extrinsic (external benefit-driven) motivations affect one's overall contributions as well as one's changes in commitment when competing demands arise. Understanding motivations of each member will help teams anticipate how members' efforts may vary within a project.

Assessing Motivations
As a team, discuss:
1. What personal satisfaction do you hope to gain from our project?
2. What tangible benefits do we hope to see come from the project?
3. What do you do when you begin to lose motivation?

> ### Principle
> The desires of the diligent will be fully satisfied, but the lazy will often be hungry.

Learning and Follow-up
Identify the principal motivations of your teammates.

How are your motivations similar or different from teammates' motivations? What might you do to address any differences?

Rate your grasp of this topic. (1=poor, 10=superb)   1 2 3 4 5 6 7 8 9 10

Date completed: Reference: Proverb 13:4

Problem Definition

Clarifying the problem being addressed and the conditions that must be met by a design solution of value to key stakeholders such as clients, investors, and the public

PROBLEM DEFINITION

| Definition | Conceptualization | Realization |

Stakeholders of Design

Context of Design
Design teams create technologies and other products which they believe people want. The needs of project clients, users, regulatory agencies, and the public shape design requirements and are used to judge acceptability and value of design solutions.

Who are your Stakeholders?
As a team, discuss:
1. Who is the primary investor in your design project?
2. Who are the likely users of your design project output?
3. Who else might be interested in what you produce?
   a. Under a best case scenario
   b. Under a worst case scenario

Principle
Choose your friends carefully so they stick with you even in times of adversity.

Learning and Follow-up
List your stakeholders and indicate their importance to your project.

How can you get to know these important individuals or groups? Why is this important to your project?

Rate your grasp of this topic. (1=poor, 10=superb)   1   2   3   4   5   6   7   8   9   10

Date completed: Reference: Proverb 18:24   15

PROBLEM DEFINITION

| Definition | Conceptualization | Realization |

Understanding Project Scope

What is Project Scope?

The scope of a project must fit the time, expertise, and resources available. Scope, defined by a client or a design team, may include:
- Objective – overall purpose of project
- Deliverables – required reports, products, approvals
- Project dates – dates for deliverables, reviews, communications
- Resources – access to funding, information, expertise
- Constraints – limits on disclosures, methods, standards, costs

Questions for Discussion

As a team, discuss:
1. What is the ideal outcome of the project?
2. What deliverables are expected by what dates?
3. What concerns or resources are off-limits to you?

Principle
You will face stern resistance when you violate a sacred boundary.

Learning and Follow-up

Create a 1-page "problem scope" document that spells out what your team should focus on achieving.

What element of the problem scope will be most limiting to your team? How do you plan to handle this?

Rate your grasp of this topic. (1=poor, 10=superb) 1 2 3 4 5 6 7 8 9 10

Date completed: Reference: Proverbs 23:10-11

PROBLEM DEFINITION

| Definition | Conceptualization | Realization |

Defining User Needs

What Does the User Want?
A project is initiated to address needs voiced by users or anticipated needs envisioned by an entrepreneur. User needs may include:
- Cost – doing something at lower cost (initial or lifetime)
- Performance – doing something better than now possible
- Responsibility – reducing dangers, waste, nuisances, pollution
- Legal – complying with laws, codes, regulations

Questions for Discussion
As a team, discuss:
1. What are desired purchase and operating costs for the solution?
2. What performances must be achieved to be competitive?
3. What legal or ethical goals need to be met?

Principle
Set your desires based on knowledge of what is good, so that hasty actions do not cause you to miss the way.

Learning and Follow-up
List specific needs you hope to meet and benefits you expect from your project solution.

Which of these may be most difficult to satisfy? Why?

Rate your grasp of this topic.
(1=poor, 10=superb)   1   2   3   4   5   6   7   8   9   10

Date completed: Reference: Proverb 19:2

PROBLEM DEFINITION

| Definition | Conceptualization | Realization |

Identifying Solution Constraints

What Constrains the Design?
A project typically faces limitations that prevent a design team from working without bounds. Project constraints may include:
- Development cost – cash flow for design, testing, certification
- Proprietary – ownership and use of patents, secrets
- Procedural – methods for design, production, distribution
- Facility – equipment, space, utility, safety limitations

Questions for Discussion
As a team, discuss:
1. What are total cost and cash flow limits for development?
2. What protection must be given to proprietary information?
3. Are organizational or industry standards to be followed?
4. What facility or equipment limitations will constrain your project?

Principle
When under someone else's authority, be careful what you do, and do not take advantage of apparent generosity.

Learning and Follow-up
List specific types of constraints that apply to your project.

How can you learn exactly what they mean to you and your project?

Rate your grasp of this topic.
(1=poor, 10=superb)   1   2   3   4   5   6   7   8   9   10

Date completed: Reference: Proverbs 23:1-3

| Definition | Conceptualization | Realization |

Information Gathering

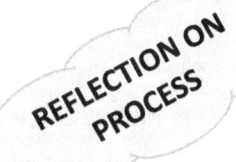

Process Steps

To this point, you have gathered information about user needs and constraints. You have taken steps that have included:
- Identifying stakeholders of the project
- Identifying project scope
- Identifying user needs for cost, performance, and responsibility
- Identifying constraints on design and development

Principle
To us, our thinking seems correct until someone else asks critical questions.

Self-Assessment

About which of the steps listed above are you most confident? Why?

For which steps do you lack user data to defend your understanding? Where can you obtain supporting data?

As necessary, go to users and obtain data to clarify their needs and constraints with certainty.

Rate your grasp of this topic.
(1=poor, 10=superb)   1   2   3   4   5   6   7   8   9   10

Date completed: Reference: Proverb 18:17

PROBLEM DEFINITION

| Definition | Conceptualization | Realization |

Value Proposition

What Justifies this Project?
A value proposition is a brief statement of value to be delivered by the solution proposed to meet a need. It includes:
- Compelling statement of the problem or opportunity
- Brief description of the envisioned solution
- Estimated value delivered to investors or users

Questions for Discussion
As a team, discuss:
1. What is the core problem being addressed?
2. What type of solution will be created to resolve the problem?
3. What major benefits will be gained by investors or users?
4. Why do you believe you can deliver value to investors?

Principle
Not delivering on what you boast is like clouds not delivering rain in a year of drought.

Learning and Follow-up
Prepare a 30-second value proposition that can "sell" your solution to the problem or opportunity at hand.

Ready yourself to present this proposition in any opportunity.

Rate your grasp of this topic.
(1=poor, 10=superb)   1   2   3   4   5   6   7   8   9   10

Date completed: Reference: Proverbs 25:14

PROBLEM DEFINITION

| Definition | Conceptualization | Realization |

Stakeholder Needs

Importance of Needs
Design efforts should focus on meeting the most important needs. Determining importance of needs requires that you consider:
- Value to principal stakeholders (client, team, users)
- Value to society and other stakeholders
- Long-term impacts on users, public, cultural norms, resources
- Achievability with team and resources available

Questions for Discussion
As a team, discuss:
1. Which are the top 5 needs? Why are they most important?
2. Which are the next 5 needs? Are they vital to project success?
3. Is anything else a vital need?

Principle
Consider not only the needs of those who are most influential but also those less influential who have real needs.

Learning and Follow-up
Tabulate user needs rated by importance (L/M/H) to the project.

Type	Description of Need	(L/M/H)
Cost	Purchase price less than _____	

Rate your grasp of this topic.
(1=poor, 10=superb)   1   2   3   4   5   6   7   8   9   10

Date completed: Reference: Proverbs 14:31

PROBLEM DEFINITION

Definition	Conceptualization	Realization

Project Constraints

Impacts of Constraints
Design constraints affect the project differently, depending if they are fixed, conditional, or negotiable:
- Fixed – must be observed, as defined, under any condition
- Conditional – limits depend upon how certain conditions are met
- Negotiable – limits can be debated and adjusted at the start

Questions for Discussion
As a team, discuss the type of constraint you face in:
1. Project deadlines and deliverables
2. Knowledge, skills, perspectives offered by team members
3. Funding available for design and development
4. Policies for documenting and protecting project assets

Principle
When you find yourself hopelessly trapped, give all diligence to pleading for freedom to do what is needed.

Learning and Follow-up
Tabulate constraints by type (fixed, conditional, negotiable):

Description of Constraint	Type Explained

Rate your grasp of this topic.
(1=poor, 10=superb)   1   2   3   4   5   6   7   8   9   10

Date completed: Reference: Proverbs 6:3-5

PROBLEM DEFINITION

| Definition | Conceptualization | Realization |

Problem Definition

What Goes in a Problem Definition?
A useful problem definition is 1 to 2 pages that concisely defines:
- Problem description and context for what is to be created
- List of user needs to be met by the solution
- List of limitations (e.g., cost, function, standards)

Questions for Discussion
As a team, discuss:
1. How might the problem definition be used by you?
2. How much detail is needed in each section of the definition?
3. What of required information have you already prepared?
4. What information yet needs to be prepared?

Principle
Give careful thought to the path forward and remain steadfast to stay on the way.

Learning and Follow-up
Prepare a Problem Definition document suitable for sharing with any of your project stakeholders.

Ask stakeholders to give you feedback on your Problem Definition to be sure you understand the problem fully.

Rate your grasp of this topic.
(1=poor, 10=superb)   1   2   3   4   5   6   7   8   9   10

Date completed: Reference: Proverb 4:26

| Definition | Conceptualization | Realization |

Defined Problem

Review Components
Compile evidence that your defined problem adequately addresses:
- Stakeholders of the project
- Project scope
- User needs for cost, performance, and responsibility
- Constraints on design and development

Assemble qualified reviewers from outside your design team.

Principle
When you do not accept knowledge and refuse expert advice, you will eat the fruit of your independent ways.

Review in Conjunction with Outside Evaluators
Rate (circle) the adequacy of:

a. Stakeholder involvement	Inadequate	Satisfactory	Excellent
b. Grasp of problem scope	Inadequate	Satisfactory	Excellent
c. Definition of user needs	Inadequate	Satisfactory	Excellent
d. Awareness of constraints	Inadequate	Satisfactory	Excellent

List elements needing further development and what is needed.

Rate adequacy to proceed: Not ready   Partially ready   Fully ready

Rate your grasp of this topic.
(1=poor, 10=superb)   1 2 3 4 5 6 7 8 9 10

Date completed: Reference: Proverb 1:28-31

Concept Generation

The search, screening, selecting, and specifying requirements for a design concept that is the foundation for a high quality design solution

CONCEPT GENERATION

| Definition | Conceptualization | Realization |

Generating Ideas

Looking for Ideas
With the problem defined, search for ideas that might develop into a full solution or part of a solution to the problem. Ideas may be original, taken from existing solutions to similar needs, or created from a combination of original and adapted solutions.

Questions for Discussion
As a team, discuss:
1. <u>Existing</u> solutions may come from familiar items, patents, product catalogs, etc. What other sources may fit your problem?
2. Generating <u>new</u> ideas may come from brainstorming, inverting ideas, or combining ideas. What other methods might be useful?
3. What components of your solution need creative ideas?

Principle
Where you search will determine the types of things you find, so choose good sources to find good ideas.

Learning and Follow-up
Assign team members to research existing solution ideas for parts of your envisioned solution. Compile ideas and sources.

Brainstorm alone, share your ideas, then brainstorm with your team. When ideas stop, brainstorm for the perfect solution, for reverse conditions, etc. Compile each idea and its source.

Rate your grasp of this topic.
(1=poor, 10=superb)   1   2   3   4   5   6   7   8   9   10

Date completed: Reference: Proverb 11:27

CONCEPT GENERATION

| Definition | Conceptualization | Realization |

Screening Ideas

Making Decisions
Large numbers of ideas are screened to identify those with greatest potential. Ideas kept for further refinement are those that best meet user needs, such as cost, functional, technical, and social concerns.

Questions for Discussion
As a team, discuss:
1. What solution costs are of greatest concern to users?
2. What solution functionality is most important to users?
3. What technical (assembly, durability, repair) issues concern users?
4. What personal or societal impacts of the solution concern users?

Principle
Remove the dross from the raw material and then the artisan can create what is valuable.

Learning and Follow-up
Use a screening matrix (shown below) to score ideas for each component or whole solution. Enter screening criteria as row labels and ideas as column labels. Score each idea (-1 = poor, 0 = ok, +1 = good) on each criterion, then total each column. Ideas with highest total scores are kept, others set aside.

Criteria	Idea 1	Idea 2	Idea 3	Idea 4
Low cost	+1	0	-1	+1
Appearance	1	-1	+1	0
Accuracy	0	0	+1	+1
Durability	-1	+1	+1	+1
TOTAL	1	0	2	3

Rate your grasp of this topic. (1=poor, 10=superb) 1 2 3 4 5 6 7 8 9 10

Date completed: Reference: Proverb 25:4

CONCEPT GENERATION

Definition	Conceptualization	Realization

Synthesizing Concepts

Identifying Feasible Concepts
A viable solution concept is a combination of component ideas that together perform all functions required in a solution. For example, an automatic bread maker must mix, shape, raise, and bake ingredients. Viable design concepts must fulfill all of these functions.

Questions for Discussion
As a team, discuss:
1. What are functions (e.g., gather, measure, detect, deliver) that your design solution must accomplish?
2. Into what categories can your generated ideas be grouped?
3. Have the ideas you generated to date addressed all essential functions? If not, you need to generate more ideas.

Principle
Do not invest in something that is unable to deliver on what is needed.

Learning and Follow-up
Use a matrix, as shown below, to identify combinations of your ideas that achieve all essential functions. Required functions are row labels. Fill in for each row all types of ideas that achieve that function. Then connect combinations that work together. Each connected set comprises a solution concept.

Function	Component Ideas			
Locate	Contact	Charge	Light	Sound
Gather	Brush	Scoop	Clamp	Suction
Transport	Throw	Slide	Bucket	
Storage	Stack	Pile	Line	

Rate your grasp of this topic. (1=poor, 10=superb)   1 2 3 4 5 6 7 8 9 10

Date completed: Reference: Proverb 17:16

CONCEPT GENERATION

| Definition | Conceptualization | Realization |

Selecting Best Concepts

Seeing to Understand
The best solution concepts meet the most important user needs, such as: desired functionality, costs, durability, serviceability, safety, and resource use. In selecting the best concepts, selection criteria are weighted to account for differences in importance.

Questions for Discussion
As a team, discuss:
1. What user needs are criteria for evaluating your concepts?
2. How would you weight these to reflect relative importance?
3. Which criteria are well-defined, and which are not yet understood?

Principle
Judge fairly, speaking up for meritorious ideas that seem not to have glamor or popular support.

Learning and Follow-up
Create a selection matrix, as shown below, for a system or the whole solution. Use weighted criteria as row labels. Column labels are concepts being considered. Score each concept by each criterion, then calculate the weighted sum for each concept. Concepts with highest weighted totals are selected.

Drive Train Criteria	Wt	Skid steer	Track	Mecanum	Omni wheel
Travel speed	3	5	2	3	4
Slope to climb	3	4	5	3	2
Turning radius	1	4	3	5	4
Durability	2	4	1	3	4
Weighted Total		39	26	29	30

Rate your grasp of this topic.
(1=poor, 10=superb)   1 2 3 4 5 6 7 8 9 10

Date completed: Reference: Proverb 31:9

| Definition | Conceptualization | Realization |

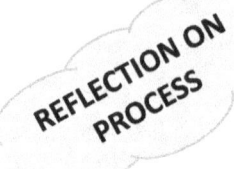

Concept Selection

Process Steps

To this point, you have selected your best concept by:
- Generating ideas from existing and original sources
- Screening ideas to meet cost, function, technical and social needs
- Synthesizing solution concepts from component ideas
- Visualizing concepts to understand how they look and work
- Selecting concepts based on weighted needs criteria

Principle
To us, our thinking seems correct until someone else asks critical questions.

Self-Assessment

About which of the steps listed above are you most confident? Why?

For which steps do you lack a strong basis for defending outcomes? How could you strengthen the step to be solid and defensible?

As necessary, go to others to clarify your understanding or obtain help to be certain about your concept selection.

Rate your grasp of this topic.
(1=poor, 10=superb)   1   2   3   4   5   6   7   8   9   10

Date completed: Reference: Proverb 18:17

CONCEPT GENERATION

| Definition | Conceptualization | Realization |

Improving the Concept

Combining Best Features
Your selected concept likely had strong and less strong scores for meeting needs. Competing concepts likely had strengths where the selected concept did not. Sometimes you can improve your solution concept by combining strengths from multiple concepts.

Questions for Discussion
As a team, discuss:
1. For what user needs did your selected concept score poorly?
2. What competing concepts scored well for meeting these needs?
3. How might you adapt strengths of other concepts to improve the selected one?

Principle
Choose the better combination, even though individual parts may be valued differently.

Learning and Follow-up
Create a selection matrix, with your selected (original) concept in column 3. Add columns with improvements on this concept, and score each by the criteria. The improved concept with the highest weighted sum becomes the newly selected concept.

Needs-Based Criteria	Wt	Original Concept	Original + Change 1	Original + Change 2
Low cost	3	4	4	5
Appearance	1	3	5	4
Accuracy	3	4	4	5
Durability	2	3	3	4
Weighted Total		**33**	**35**	**42**

Rate your grasp of this topic.
(1=poor, 10=superb)   1 2 3 4 5 6 7 8 9 10

Date completed: Reference: Proverb 15:17

CONCEPT GENERATION

Definition	Conceptualization	Realization

Specifying Component Interfaces

Identifying Interfaces between Components
A design solution typically is comprised of major components or systems. These component parts must interact by sharing space or by transfering material, signals, power, or control. At the concept stage of design, component interfaces need to be specified.

Questions for Discussion
As a team, discuss:
1. What space sharing or positioning is required among parts?
2. What limits are required for material passing between parts?
3. What power must be transferred between parts?
4. What information, signal, data transfer must occur among parts?

Principle
When one part gives freely what is needed by another, the giver benefits as well as the receiver.

Learning and Follow-up
List the interface requirements for your solution, such as:

Type	Description
Material	Collected berries must be transferred individually to the color sensor
Signal	Color signal must be transferred to the sorter gate

Rate your grasp of this topic.
(1=poor, 10=superb)   1 2 3 4 5 6 7 8 9 10

Date completed: Reference: Proverb 11:24

CONCEPT GENERATION

Definition	Conceptualization	Realization

Specifying Functionality

Identifying Functional Requirements
A successful design solution must provide desired functionality and perform to levels expected by stakeholders. Functional requirements need to be specific to the concept that was selected so they guide decisions as the concept becomes a fully developed solution.

Questions for Discussion
As a team, discuss:
1. What are speed or capacity requirements for the solution?
2. What accuracy or repeatability requirements are important?
3. What space or reach or confinement limits are crucial?
4. What other functionality is important for the solution?

Principle
Define clear measures for what is desired, and do not change them unfairly when evaluating your design in the future.

Learning and Follow-up
List and specify values for important functional requirements for your envisioned solution. Be as specific as possible.

Description	Value
Time to complete 1 cycle	3 seconds

Rate your grasp of this topic. (1=poor, 10=superb)   1 2 3 4 5 6 7 8 9 10

Date completed: Reference: Proverb 20:23

CONCEPT GENERATION

Definition	Conceptualization	Realization

Specifying Costs

Identifying Financial Requirements
Once the solution concept is selected, measurable requirements need to be established to evaluate the solution as it is developed. Among financial requirements may be materials cost, purchase price, operating cost, and life cycle costs.

Questions for Discussion
As a team, discuss any required cost limits for:
1. Materials used to make the product
2. Sale price, packaging and delivery, or user purchase price
3. Consumables, utilities, servicing, repairs for normal operation
4. Installation, certification, upgrades, insurance, disposal

Principle
A person begrudging or uncertain about costs may restrict unnecessarily expenses that are reasonable.

Learning and Follow-up
List and specify financial requirements for your envisioned solution. Be as specific as possible at this stage of the project.

Cost type	Description	Limit
Cost of materials	Parts, cartridges, paint	<$1250

Rate your grasp of this topic. (1=poor, 10=superb)   1 2 3 4 5 6 7 8 9 10

Date completed: Reference: Proverb 23:7

CONCEPT GENERATION

Definition	Conceptualization	Realization

Specifying Technical Requirements

Identifying Technical Requirements
Some solutions must meet technical specifications, industry standards, or certified performances. Such requirements may specify material properties, testing procedures, manufacturing or repair practices, communication firewalls, or parts interchangeability.

Questions for Discussion
As a team, discuss any:
1. Industry or government standards or regulations that apply
2. Testing or certifications required
3. Requirements for manufacture or repair or disposal
4. Security or trade secrets to be protected

Principle
Know and follow the wisdom of those who have gone before you and you will be safe and not stumble.

Learning and Follow-up
Specify technical requirements appropriate for your envisioned solution. Be as specific as possible at this stage of the project.

Description	Standard or Limit
Soldering electrical connections	IPC J-STD-001
Maximum replacement time for motor	2 hours

Rate your grasp of this topic.
(1=poor, 10=superb)   1   2   3   4   5   6   7   8   9   10

Date completed: Reference: Proverb 21:23

CONCEPT GENERATION

| Definition | Conceptualization | Realization |

Specifying Social Requirements

Identifying Social Requirements
Design solutions must not subject workers, bystanders, people groups, or society in general to unreasonable risks or disadvantages. This requires consideration of direct impacts from technologies as well as more subtle favoritism and empowerment over time.

Questions for Discussion
As a team, discuss any concerns about your solution's:
1. Noise, vibration, moving parts, potential eruptions
2. Harmful radiation, chemical release, infectious organisms
3. Economic, social impacts on individuals, communities, or cultures
4. Degradation of environment, natural resources, organisms

Principle
Pay attention to and actively care for those who are or who may become disadvantaged through your actions.

Learning and Follow-up
Specify requirements that provide protections for those who might be affected by your design solution over time.

Requirement Description	Specification
Maximum noise level for user	80 dB at 1 ft distance
Design to prevent safety hazards	NIOSH PtD practices

Rate your grasp of this topic.
(1=poor, 10=superb)   1   2   3   4   5   6   7   8   9   10

Date completed: Reference: Proverb 29:7

| Definition | Conceptualization | Realization |

Selected Concept

Review Components
Compile evidence that your selected concept adequately addresses:
- Functional requirements of the project
- Financial requirements
- Technical requirements
- Social requirements

Assemble qualified reviewers from outside your design team.

Principle
When you do not accept knowledge and refuse expert advice, you will eat the fruit of your independent ways.

Review in Conjunction with Outside Evaluators
Rate (circle) the adequacy of your concept's potential to meet:

a. Functional requirements	*Inadequate*	*Satisfactory*	*Excellent*
b. Financial requirements	*Inadequate*	*Satisfactory*	*Excellent*
c. Technical requirements	*Inadequate*	*Satisfactory*	*Excellent*
d. Social requirements	*Inadequate*	*Satisfactory*	*Excellent*

List elements of the concept or requirements needing further development. Identify what action is needed.

Rate adequacy to proceed: *Not ready   Partially ready   Fully ready*

Rate your grasp of this topic. (1=poor, 10=superb)   1 2 3 4 5 6 7 8 9 10

Date completed: Reference: Proverb 1:28-31

Solution Completion

The development of a detailed solution -- from a solution concept to one that is usable or ready for hand-off for additional testing, refinement, or implementation

SOLUTION COMPLETION

Definition	Conceptualization	Realization

Design for Prevention

Identifying Potential Failures
All technologies have the potential to fail, and some failures will have serious impacts on cost, safety, or performance. Preventative design identifies potential failures and designs them out of the solution. Risk of failure is quantified by a risk factor, R: $R = L * S$, where
L = likelihood of failure, S = seriousness of failure.

Questions for Discussion
As a team, identify for your envisioned design solution:
1. Failures that can cause safety hazards
2. Failures that can reduce how well the solution performs
3. Design flaws that require costly repairs or time loss
4. Design failures that could cause rejection of the solution

Principle
When you know a path that produces undesirable results, avoid it at all costs.

Learning and Follow-up
Create a table for calculating risk factors, as defined below:
Likelihood: 1 = very low, 5 = medium, 10 = high, 100 = very high
Seriousness: 1 = negligible, 5 = minor, 10 = serious, 100 = fatal

Type of Failure	L	S	R
Conveyor chain breaks during use	3	7	21

Make design changes necessary to reduce risk factors below 10.

Rate your grasp of this topic.
(1=poor, 10=superb)   1 2 3 4 5 6 7 8 9 10

Date completed: Reference: Proverb 16:17

SOLUTION COMPLETION

| Definition | Conceptualization | Realization |

Design for Efficiency

Modeling and Analysis
Engineering analysis and modeling enable us to predict results before we complete our design. With mathematical models, visualization software, physical mock-ups, and sketches, we can predict at what point a failure might occur, how well a component might perform, costs of design alternatives, or how a product might look.

Questions for Discussion
As a team, identify where analysis or modeling can be useful in:
1. Estimating costs of different design alternatives
2. Predicting how the design will perform under different conditions
3. Sizing or selecting components for durability
4. Visualizing features of the design

Principle
One answer may seem right until someone probes it to really understand what is correct.

Learning and Follow-up
Compile a list of questions that need to be probed, and identify a method of analysis or modeling that can be used for each.

Design Question	Method of Analysis
What steel plate thickness will carry the load?	CAD and finite element analysis
What is the rate of return on replacing the press with model FP-106?	Life cycle cost analysis

Rate your grasp of this topic. (1=poor, 10=superb)   1 2 3 4 5 6 7 8 9 10

Date completed: Reference: Proverb 18:17

SOLUTION COMPLETION

Definition	Conceptualization	Realization

Design for Producability

Assembly and Repair
Before a design is finished, we must ensure that it can be produced and maintained. We must be careful to choose, order, and link components of the design solution so they can be easily assembled, serviced, upgraded, and replaced.

Questions for Discussion
As a team, identify what is required for:
1. Components to be assembled to implement the solution
2. Regular servicing or preventative maintenance that is necessary
3. Periodic upgrades or replacement of components
4. Repairs or replacement when serious problems occur

Principle
Enabling what is hoped for brings joy, but delaying action and not delivering on the hope brings discouragement.

Learning and Follow-up
Compile a list of assembly or servicing issues and identify what design features may be desired to address the need.

Assembly or Servicing Need	Design to Address Need
Replacement of worn chain	Removable guards, adjustable tighteners
Periodic software fixes or upgrades	Remote access and two-level security
Ease in replacing parts	Use standard parts

Rate your grasp of this topic.
(1=poor, 10=superb)   1 2 3 4 5 6 7 8 9 10

Date completed: Reference: Proverb 13:12

SOLUTION COMPLETION

Definition	Conceptualization	Realization

Design for Sustainability

Sustainability Issues
Being socially responsible in design requires us to follow sustainable design practices. Sustainable design seeks to reduce consumption of non-renewable resources, minimize waste, and create healthy environments.

Questions for Discussion
As a team, discuss:
1. Materials used, their renewability, and polluting characteristics
2. Manufacturing processes, energy use and waste generated
3. Impact on people and the environment when the solution is implemented
4. Life cycle costs of sustainable vs. non sustainable alternatives

Principle
Give careful thought to your ways, staying away from evil, and be focused in follow-through to reach your goal.

Learning and Follow-up
List issues of sustainability in your design solution and identify ways to address each issue.

Sustainability Issue	Ways to Address Issue
Consumption of rare metals	Choose sensors not using rare metals
Water pollution from cleanup	Waste reservoir and water recycling

Rate your grasp of this topic. (1=poor, 10=superb)   1 2 3 4 5 6 7 8 9 10

Date completed: Reference: Proverb 4:26

SOLUTION COMPLETION
↓

| Definition | Conceptualization | Realization |

Prototyping for Component Testing

Planning for Prototype Testing
Prototyping is useful for early testing of designs. By testing parts or subsystems, problems can be identified and remedied before they are masked by solution complexity. A useful prototype must provide functionality or features being evaluated, but need not fully model the solution. Prototypes should enable quick, cost-effective testing.

Questions for Discussion
As a team, identify design features that might be prototyped:
1. Subsystems or components that should be tested separately
2. Types of prototype that best model desired features
3. Resources and expertise needed to do this prototyping

Principle
To gain wisdom and understanding, carefully listen, watch, and wait to fully grasp what can be learned.

Learning and Follow-up
Identify solution components that might best be prototyped to enable convenient and cost-effective testing.

Feature or Function	Type of Prototype
Robot arm reach and grip of object	Erector set assembly of arm components and actuators
Reaction time for location sensing and movement to desired position	Circuit mock-up with sensors and actuators

Rate your grasp of this topic.
(1=poor, 10=superb)   1   2   3   4   5   6   7   8   9   10

Date completed: Reference: Proverb 8:34

SOLUTION COMPLETION

| Definition | Conceptualization | Realization |

Testing and Evaluation

Testing Principles

Testing of a design solution, all or in part, can measure performance, identify failures, or evaluate durability. Test procedures and data analysis will depend upon questions asked:
a. *How much, how consistent*: Average values, standard deviations
b. *Effects of conditions*: Charts, equations, analysis of variance
c. *How frequently*: Frequency plots

Questions for Discussion

As a team, identify important performances that call for testing:
1. Performance levels to be met or exceeded
2. Frequency of failures of different types
3. Consistency of performances under varied conditions

Principle
The person who evaluates and judges fairly gains credibility so that others listen and follow.

Learning and Follow-up

Create a test plan, such as below, for testing the design solution.

Test	Test Plan
Required placement of object in target within 5 seconds	1. Actuate mechanism with object in hand 2. Measure time to deliver object 3. Repeat, record time for 10 identical tests 4. Calculate mean time to deliver object
Determine variability in task completion with different operators	1. Define task to be done and conditions 2. Operators, 3 times each in random order, complete task; record completion time 3. Calculate average time for each operator 4. Use T-statistic to test if differences are significant

Rate your grasp of this topic.
(1=poor, 10=superb)   1   2   3   4   5   6   7   8   9   10

Date completed: Reference: Proverb 29:14

SOLUTION COMPLETION

Definition	Conceptualization	Realization

Iterating to Revisit Earlier Work

Iteration
At times, new information reveals that previous work was not done adequately. Iteration, returning to repeat earlier work, is a frequent part of design and essential to quality design. Examples include:
a. Redefining user needs because a key stakeholder is now heard
b. Considering new concepts after a selected one is found lacking
c. Revising a requirement after solution testing shows it in error

Questions for Discussion
As a team, discuss:
1. Where has your team used iteration? Did it improve your design?
2. Where in your design work do you see inadequacies that may require returning to revise earlier work? What might be gained?

Principle
When we turn around and make changes in the face of being corrected, we open ourselves to great learning.

Learning and Follow-up
Tabulate instances where iteration was used or is needed, such as:

Need for Iteration	Result of Iteration
The selected and prototyped gathering mechanism was unable to achieve its targeted gathering rate.	A new concept was identified, investigated, and prototyped, and its gathering rate was compared to the original design. Testing and revision of the new concept doubled the gathering rate for the overall system.
Gyroscopic sensor has proven unable to provide accurate turn measurements.	Plans are to test sensor B as a possible replacement for sensor A. Both sensors will need to be tested under operating conditions simulating expected applications.

Rate your grasp of this topic.
(1=poor, 10=superb)   1 2 3 4 5 6 7 8 9 10

Date completed: Reference: Proverb 1:23

| Definition | Conceptualization | Realization |

Solution Completion

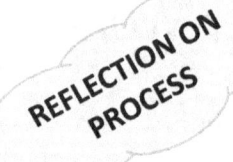

Process Steps

To this point, you have developed and refined your solution by:
- Designing to lower risks of failures
- Modeling and analyzing to improve efficiency
- Specifying materials and processes that are sustainable
- Creating prototypes that enable component testing
- Conducting tests to evaluate performance and affect design revisions

Principle

To us, our thinking seems correct until someone else asks critical questions.

Self-Assessment

About which of the steps listed above are you most confident? Why?

For which steps do you lack a strong basis for defending outcomes? How could you strengthen the step to be solid and defensible?

As necessary, go to others to clarify your understanding or obtain help to be certain about your solution development.

Rate your grasp of this topic. (1=poor, 10=superb)   1 2 3 4 5 6 7 8 9 10

Date completed: Reference: Proverb 18:17

SOLUTION COMPLETION ↓

| Definition | Conceptualization | Realization |

Verifying Solution

Verification of Expectations
Solution verification determines how well the solution meets requirements against which it was developed. If requirements cannot be met, you may need to gather necessary data, make design changes to improve performance, or eliminate or revise problematic requirements.

Questions for Discussion
As a team, discuss how you will validate your solution:
1. Which requirements must you prove to have met?
2. What data is available? What is yet needed?
3. How will you build a solid case for solution validation?

Principle
When you falsely claim not to know of a problem, you will eventually be repaid for what you have done.

Learning and Follow-up
Create a verification plan, such as below, for solution requirements.

Requirement	Status	Yet to be Done
Cycle time < 5 sec	Cycle time 6.5 sec	Redesign actuator
Better balance than XYZ device	Not tested	Market study with users
Time between failures >1000 hours	Appears to be 200 hours	Reduce target to 500 hours, specify maintenance at 100 hour intervals

If verification reveals significant inadequacies, return to earlier steps to modify requirements or to make needed design changes.

Rate your grasp of this topic. (1=poor, 10=superb)   1 2 3 4 5 6 7 8 9 10

Date completed: Reference: Proverb 24:12

SOLUTION COMPLETION

| Definition | Conceptualization | **Realization** |

Refining Solution

Refinement of Solution
When the solution meets requirements, minor refinements can yet be made to make it more attractive, more intuitive, more durable, or more acceptable amidst competitors. As delivery dates approach, changes should not add new risks nor weaken proven strengths.

Questions for Discussion
As a team, discuss what refinements are justified at this time:
1. Replacing custom parts with standard parts
2. Making assembly, repairs, and updates easier to accomplish
3. Enhancing appearance, feel, or enjoyment to users
4. Making more intuitive, reducing training and need for instruction
5. Using materials or processes that are more sustainable

Principle
The patient person with self-control will produce better results than the warrior fighting to take control.

Learning and Follow-up
Identify and prioritize refinements to your solution based on potential benefits and risks.

Refinement	Benefits	Risks
Replace belt drive by chain drive	More durable, less slippage	Not tested, balance affected
Replace wiring sleeves by carriers	Movement less restricted	Increased cost, needs more space

Rate your grasp of this topic. (1=poor, 10=superb) 1 2 3 4 5 6 7 8 9 10

Date completed: Reference: Proverb 16:32

SOLUTION COMPLETION

| Definition | Conceptualization | Realization |

Validating Solution

Validation of Solution
The bottom line in design is satisfying the customer (user, investor, others directly affected). Validation is checking that key stakeholders are satisfied where they use the solution. Validation can be done by:
- Inviting potential users to try out the solution
- Conducting market research to see how potential buyers react
- Obtaining feedback from community members

Questions for Discussion
As a team, discuss what validation steps fit your solution:
1. User feedback
2. Buyer feedback
3. Community feedback

Principle
The happy person finds energy and satisfaction, while the disappointed experiences dryness in spirit.

Learning and Follow-up
Create a plan that accurately validates your solution.

Stakeholder	Validation
User (college students)	Focus group of students to see what they like and dislike about the solution
User (homemakers)	Give free product to sample of homemakers, asking for feedback in return

If feedback shows poor satisfaction with the design product, identify issues and return to earlier steps to make changes to the design.

Rate your grasp of this topic. (1=poor, 10=superb)   1 2 3 4 5 6 7 8 9 10

Date completed: Reference: Proverb 115:13

| Definition | Conceptualization | Realization |

Completed Solution

Review Components

Compile evidence that your design solution adequately meets:
- Functional requirements of the project
- Financial requirements
- Technical requirements
- Social requirements

Assemble qualified reviewers from outside your design team.

Principle
When you do not accept knowledge and refuse expert advice, you will eat the fruit of your independent ways.

Review in Conjunction with Outside Evaluators

Rate (circle) the adequacy of your solution's achievement of:
- a. Functional requirements   *Inadequate*   *Satisfactory*   *Excellent*
- b. Financial requirements   *Inadequate*   *Satisfactory*   *Excellent*
- c. Technical requirements   *Inadequate*   *Satisfactory*   *Excellent*
- d. Social requirements   *Inadequate*   *Satisfactory*   *Excellent*

List elements of the solution that are proven to be noteworthy.

Rate adequacy for handoff: *Not ready*   *Partially ready*   *Fully ready*

Rate your grasp of this topic.
(1=poor, 10=superb)   1   2   3   4   5   6   7   8   9   10

Date completed: Reference: Proverb 1:28-31

★★★★★ DESIGN THINKING ACHIEVED ★★★★★

Provide a frank, thoughtful assessment of your design thinking. Rate yourself (1 low to 5 high) on the following:

Items

Item	Poor				Superb
Researching problem/opportunity for understanding	1	2	3	4	5
Defining needs and wants of stakeholders	1	2	3	4	5
Generating innovative ideas to fuel solution concepts	1	2	3	4	5
Synthesizing solution concepts with potential value	1	2	3	4	5
Selecting concepts with most potential to meet needs	1	2	3	4	5
Identifying best design foci for adding value to solution	1	2	3	4	5
Using modeling and analysis to aid design decisions	1	2	3	4	5
Testing and evaluating design elements	1	2	3	4	5
Iterating to revisit and improve earlier design work	1	2	3	4	5
Verifying achievement of solution requirements	1	2	3	4	5
Refining solution to add value and meet needs	1	2	3	4	5
Validating that solution meets stakeholder needs	1	2	3	4	5

Principle: Be sure you periodically check on the condition of things for which you are responsible.

Think about this

1. How does your design performance (items above) compare with what it should be?

2. What might you do in the future to build on strengths or address concerns?

Reference: Proverb 27:23

★★★★★ DESIGN ★★★★★
LEARNING ACHIEVED

For each area below, rate how much you learned in your design experience. Then give an example to support ranking.

What I learned from my project experience:	None				Extensive
How to develop a useful problem definition	1	2	3	4	5
How to generate valuable solution concepts	1	2	3	4	5
How to select concepts of greatest solution merit	1	2	3	4	5
How to develop concepts into valuable solutions	1	2	3	4	5
How determine how well a solution meets real needs	1	2	3	4	5

Principle: Apply your heart to what you observed and you will learn important life lessons.

Think about this
1. How does your learning compare with what it should be from a design project experience?

2. What might you do in the future to learn more from such experiences?

Reference: Proverb 24:32

www.ingramcontent.com/pod-product-compliance
Lightning Source LLC
Chambersburg PA
CBHW030054230526
45471CB00003B/1088